| DATE DUE | | | |
|---|---|---|---|
| | | | |
| | | | |
| | | | |
| | | | |
| | | | |
| | | | |
| | | | |
| | | | |
| | | | |
| | | | |
| | | | |
| | | | |
| | | | |
| | | | |

8466

589.3 Greenaway, Theresa.
GRE
  The first plants.

**MESA VERDE MIDDLE SCHOOL**
**POWAY UNIFIED SCHOOL DISTRICT**

378867 01569 52282A 02672E    15

### DECLARATION

I hereby declare that
all the paper produced
by Cartiere del Garda S.p.A.
in its Riva del Garda mill
is manufactured completely
<u>Acid-free and Wood-free</u>

Dr. Alois Lueftinger
Managing Director and General Manager
Cartiere del Garda S.p.A.

# GREEN WORLD

# THE FIRST PLANTS

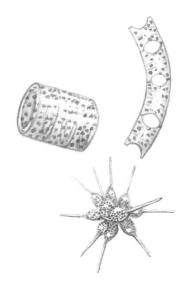

Written by
**Theresa Greenaway**

STECK-VAUGHN
L I B R A R Y
A Division of Steck-Vaughn Company

Austin, Texas

Published in the United States in 1991
by Steck-Vaughn, Co., Austin, Texas,
a subsidiary of National Education Corporation

**A Templar Book**
Devised and produced by The Templar Company plc
Pippbrook Mill, London Road, Dorking, Surrey RH4 1JE, Great Britain
Copyright © 1990 by The Templar Company plc
All rights reserved. No part of this publication may be reproduced, stored in a retrieval system, or transmitted in any form or by any means, electronic, mechanical, photocopying, recording or otherwise, without the prior permission of the publishers or copyright holders.

*Editor:* Wendy Madgwick
*Designer:* Jane Hunt
*Illustrator:* Rosie Vane-Wright

**Notes to Reader**
There are some words in this book that are printed in **bold** type. A brief explanation of these words is given in the glossary on p. 44.

All living things are given two Latin names when first classified by a scientist. Some of them also have a common name, for example giant kelp, *Macrocystis pyrifera*. In this book, the common name is used where possible, but the scientific name is given when first mentioned.

**Library of Congress Cataloging-in-Publication Data**
Greenaway, Theresa, 1947-
First plants / written by Theresa Greenaway.
p. cm. – (The Green World)
"A Templar Book" – T.p. verso.
Includes bibliographical references and index.
Summary: Discusses different types of algae and the uses of algae as a food crop and energy source.
ISBN 0-8114-2734-X
1. Algae – Juvenile literature.  [1. Algae.]  I. Title.  II. Series.
QK566.5.G74    1991    90-10003
589.3–dc20              CIP AC

Color separations by Positive Colour Ltd, Maldon, Essex, Great Britain
Printed and bound by L.E.G.O., Vicenza, Italy
1 2 3 4 5 6 7 8 9 0  LE 95 94 93 92 91

**Photographic credits**
*t = top, b = bottom, l = left, r = right*
Cover: Planet Earth; page 9 Bruce Coleman/K. Taylor; page 13 Frank Lane/Eric and David Hosking; page 15 Bruce Coleman/N. Decore III; page 17 Frank Lane/Silvestris; page 19t Frank lane/Eric and David Hosking; page 19b Bruce Coleman/M.T. O'Keefe; page 20 Frank Lane/R. Tidman; page 21 Bruce Coleman/R. Tidman; page 25 Frank Lane/Holt Studios; page 29 Bruce Coleman/M. Fogden; page 30 Planet Earth/M. Coleman; page 33t Bruce Coleman/J. Burton; page 33b Bruce Coleman/G.D. Plage; page 35 Audience Planners; page 37t Frank Lane/Eric and David Hosking; page 37b Frank Lane; page 38 Environmental Picture Library/Paul Ferraby; page 39 Bruce Coleman/M.T. O'Keefe; page 41 G.S.F. Picture Library; page 42 Dr D. Hall, Kings College, London University; page 43 Bruce Coleman.

# CONTENTS

Green World ............................................................ 6
Algae ....................................................................... 8
Different Kinds of Algae ..................................... 10
The Open Ocean .................................................. 12
Temperate Shores ................................................ 14
Rocky Tropical Shores ........................................ 16
Lakes ..................................................................... 18
Algae of Other Habitats ...................................... 20
Shape and Structure ............................................ 22
Movement and Growth ....................................... 24
Life Cycles of Algae ............................................. 26
Grass of the Sea ................................................... 28
The Kelp Forest ................................................... 30
Life in Fresh Water ............................................. 32
Threats to Lakes .................................................. 34
Threats to Marine Life ........................................ 36
Reservoirs ............................................................. 38
Seaweed and Industry ......................................... 40
The Future ............................................................ 42
Glossary ................................................................ 44
Further Reading .................................................. 44
Algae in This Book .............................................. 45
Index ..................................................................... 46

# GREEN WORLD

This tree shows the different groups of plants that are found in the world. It does not show how they developed or their relationship with each other.

**Blue-green algae**
(Cyanophyta)
- These are sometimes classed as bacteria
- Very small, with primitive cells

**Euglenas**
(Euglenophyta)
- Single-celled algae with flagella and a flexible "skin"

**Dinoflagellates**
(Dinophyta)
- Single-celled, with flagella and reddish pigment

**Golden-brown algae**
(Chrysophyta)
- Mostly single-celled
- Includes diatoms

**Green algae**
(Chlorophyta)
- Mostly freshwater, but some are seaweeds
- Chlorophyll is not hidden by other pigments. Includes stoneworts

**Brown algae**
(Phaeophyta)
- All contain brown pigment
- Mostly seaweeds – kelps, wracks, thongweeds

**Red algae**
(Rhodophyta)
- Red seaweeds and a few freshwater algae
- Contain red pigment (occasionally blue pigment)

The land area of the world is divided into ten main zones depending on the plants that grow there. Algae are found throughout the world, even on the snow and frozen rocks.

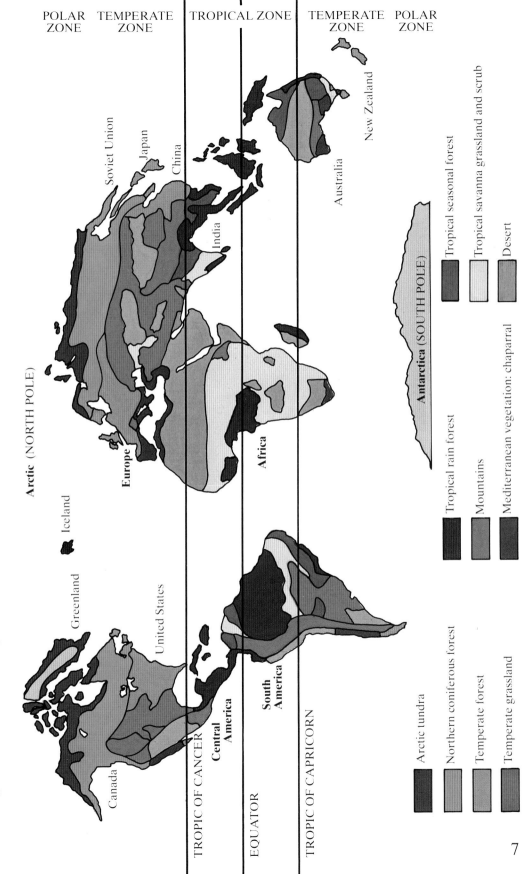

# ALGAE

Algae are the simplest plants. They live in water, soil, and other damp places. An alga may consist of only one or a few cells. Some kinds are made up of a long line of cells, which form a long, very thin thread called a **filament**. They range from tiny plants too small to see to large seaweeds.

Like land plants, all algae contain a green pigment called **chlorophyll**. Chlorophyll uses the energy in sunlight to combine carbon dioxide gas and water to make the sugars that are the plant's food by a process called **photosynthesis** (see p. 22)

Each group of algae also has its own particular color, which is produced by one or more pigments. The green color of the chlorophyll is often hidden by a red, brown, or yellow pigment.

### All sizes and shapes

The smaller algae cannot be seen at all with the naked eye, and a microscope is used to magnify them. Many live in the sea. The largest and most complex algae also live in the sea – they are better known as seaweeds. The largest one of all is a brown seaweed called giant kelp (*Macrocystis pyrifera*), which grows up to 400 feet long. Even these large algae do not have roots, leaves, flowers, or fruits like flowering plants.

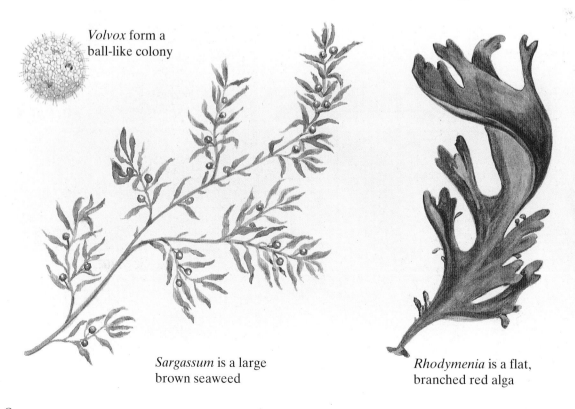

*Volvox* form a ball-like colony

*Sargassum* is a large brown seaweed

*Rhodymenia* is a flat, branched red alga

**Earliest plants**

The Earth formed about 4.5 billion years ago. As the crust of the Earth cooled, gases from the hot interior made the first atmosphere. It was different from the air today, with carbon dioxide, ammonia, and methane gases mixed with clouds of water vapor but no oxygen.

The earliest signs of life date from about 3 billion years ago. Blue-green algae were the very first green plants. They lived in the bottoms of pools and lakes, where the water screened them from the sun's lethal ultraviolet rays. They multiplied and spread, and new groups appeared. The oxygen they produced during photosynthesis bubbled up through the water and into the air. As the amount of oxygen increased, it stopped more and more harmful ultraviolet light from reaching Earth. Eventually, plants and animals could leave the protection of the water and live safely on land.

Knowledge about the origins of the first plants comes from carefully studying the fossils in some of the world's oldest rocks. Some of the blue-green algae alive today are surprisingly similar to their ancient ancestors.

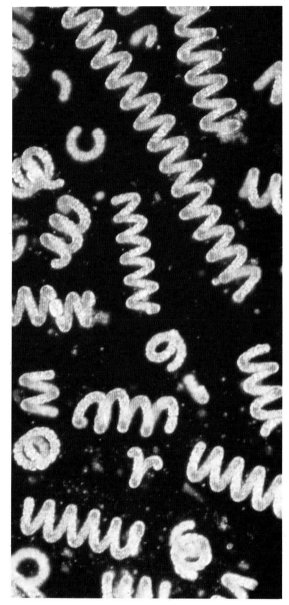

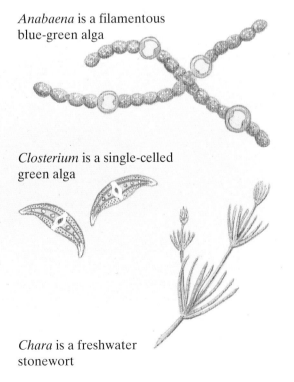

*Anabaena* is a filamentous blue-green alga

*Closterium* is a single-celled green alga

*Chara* is a freshwater stonewort

- Algae grow in water, soil, and damp places.
- All algae contain chlorophyll and make their own food by photosynthesis.
- One-celled, or unicellular, algae are very small. Some are less than 10 micrometers. (There are about 4,000 micrometers in an inch.)
- Other algae are large and complex, but they do not have roots, stems, or leaves.
- Many algae contain colored pigments.

# DIFFERENT KINDS OF ALGAE

There are many types of algae. The simplest and most ancient kind are the blue-green algae. There are about 2,000 species and their cells differ from those of all other algae. They are similar to bacteria and so they are sometimes called blue-green bacteria. However, as they contain chlorophyll they are usually placed with algae.

The largest algae are the seaweeds which can be divided into three groups according to their color: red, brown, or green. Some green algae called stoneworts are put in a group of their own. There are about 300 species, and they grow in ponds and streams with muddy bottoms. Some have a hard outer layer and can be up to three feet in height.

The other groups of algae are all single cells or small filaments. Some groups, in particular dinoflagellates and diatoms, make up most of the marine and freshwater **plankton** that float at the surface of the water.

**Blue-green algae**
There are three kinds of blue-green algae: single-celled forms; filamentous (made of filaments); or colonial (groups of cells). They are often surrounded by jelly and the jellylike blobs can become quite large, although the individual plants are tiny.

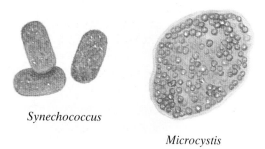

*Synechococcus*

*Microcystis*

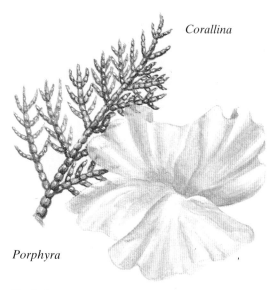

*Corallina*

*Porphyra*

*Oscillatoria*

**Red algae**
Most of the 4,000 species of red algae are marine (live in the ocean). Only a few kinds live in fresh water. Some red seaweeds encrust rocks and larger seaweeds, while others grow into branched fronds. They may be smooth to touch, or stiff and rough with a chalky layer made by the plant. These algae are found along shores worldwide, but most grow on tropical coastlines.

## Brown algae

The 2,000 species of brown algae are nearly all marine. There are no single-celled kinds – the smallest brown algae are tiny, branched filaments. The rest are the mostly large brown seaweeds. They are an important part of the marine vegetation of cooler coastlines.

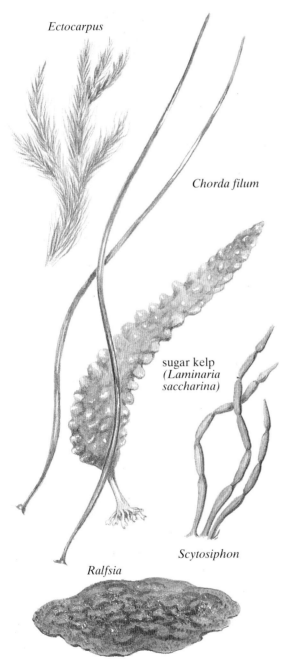

*Ectocarpus*

*Chorda filum*

sugar kelp (*Laminaria saccharina*)

*Scytosiphon*

*Ralfsia*

## Green algae

There are about 10,000 species of green algae, which range in size from single-celled kinds to medium-sized green seaweeds. The largest green algae are marine forms, but 90 percent of this group live in fresh water. The green algae have a very wide range of shapes.

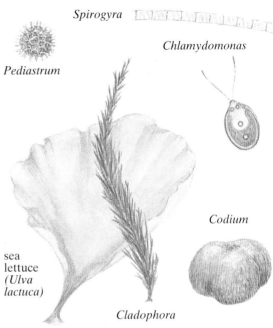

*Spirogyra*

*Chlamydomonas*

*Pediastrum*

sea lettuce (*Ulva lactuca*)

*Codium*

*Cladophora*

## Diatoms

There are about 10,000 species of diatoms. They all have a "shell" of silica which is in two parts, like a box and lid. Though so tiny (just 20–200 micrometers long), the shell is often finely patterned. Dinoflagellates, euglenas, and cryptomonads possess tiny hairs, or **flagella**. By lashing these they move feebly through water.

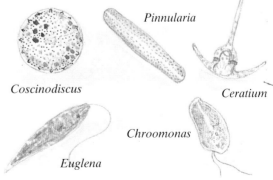

*Pinnularia*

*Coscinodiscus*

*Ceratium*

*Chroomonas*

*Euglena*

# THE OPEN OCEAN

Sailors and scientists once thought that there were no plants far out in the ocean. They could see fishes, whales, and jellyfish. They did not know that the oceans were teeming with the microscopic plants and animals that are now called plankton, which drift on the surface of open waters.

In the 100 years or so since their discovery, scientists have come to realize the importance of these tiny algae. The oceans cover 70 percent of the world's surface. Planktonic algae are found from the shallow waters along the coasts to the middle of the oceans, far out of sight of land. There are so very many of these minute plants that they have been called "the grass of the sea." This is a good description, for just as grass supports food chains of animals on the land, so these algae are at the bottom of the marine food chains.

### Collecting plankton

To find out about planktonic algae, small amounts of seawater called samples are collected. These samples are taken to a laboratory and examined under a microscope. Different depths of water are sampled by using special kinds of plankton nets. There are greater numbers of planktonic algae in colder waters than in the warm, tropical seas near the equator. Most of these algae belong to the diatom and dinoflagellate groups.

The average depth of the oceans is 12,500 feet. The sun's rays penetrate to a depth of 1,500 feet, but 80 percent of the light is filtered out in the top 35 feet. As planktonic algae need enough light for photosynthesis, they can only live in the top 50 feet of the ocean.

## Diatoms

Diatoms are numerous in colder waters. It is difficult to find out how they live because they are so small. Marine diatoms are mostly circular species and they often join to form chains. Even though their silica shells are tiny, they are heavier than seawater. No one is quite sure how they remain floating near the surface.

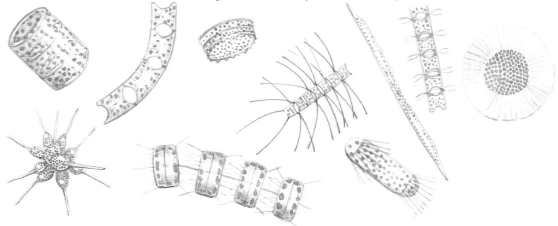

## Dinoflagellates

Many of these curiously shaped plants have hairs to propel them through the water. They may reach speeds of up to 1 inch a minute. They are interesting for another reason – they can make their own flashes of light. This light is called **bioluminescence**. It can be seen at night and may be set off by water movement.

Dinoflagellates are a widespread group and are found in all kinds of waters. Some species prefer warmer regions. Occasionally their numbers increase very rapidly and when this happens they color the water red because of the pigment they contain. Unfortunately, some dinoflagellates are poisonous to fish. A species called *Gymnodinium brevis* has killed thousands of fish in such outbreaks. Another microscopic dinoflagellate, *Gonyaulax*, collects inside shellfish. It does not harm them, but can poison and kill people who eat the shellfish.

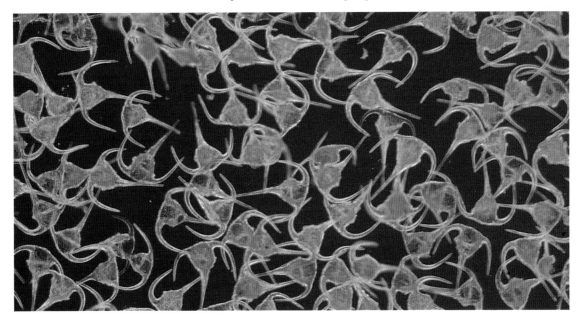

# TEMPERATE SHORES

All around the world the sea flows and retreats up and down the beaches twice a day. The area over which the tide sweeps is called the **littoral zone**. The littoral zone is divided into the upper, middle, and lower shore. Different kinds of seaweed grow at each level. They have to be tough enough to withstand often-violent waves. Few seaweeds are found on soft sands or a stony beach where the pebbles are rolled to and fro.

Along the coasts of the North Atlantic Ocean (North America, Iceland, Scandinavia, Britain, and Europe) the seaweeds of the littoral zone have a similar distribution. On the Atlantic coasts of the southern United States, *Sargassum*, a large brown wrack, grows on the lower shore. Fronds torn off by strong waves drift to the Sargasso Sea, an area of the North Atlantic southeast of Bermuda. The floating fronds grow well there but cannot reproduce. Eventually, they become so large that they sink.

**The littoral zone**

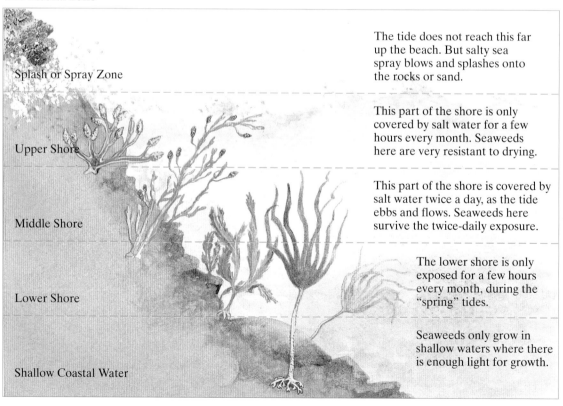

Splash or Spray Zone — The tide does not reach this far up the beach. But salty sea spray blows and splashes onto the rocks or sand.

Upper Shore — This part of the shore is only covered by salt water for a few hours every month. Seaweeds here are very resistant to drying.

Middle Shore — This part of the shore is covered by salt water twice a day, as the tide ebbs and flows. Seaweeds here survive the twice-daily exposure.

Lower Shore — The lower shore is only exposed for a few hours every month, during the "spring" tides.

Shallow Coastal Water — Seaweeds only grow in shallow waters where there is enough light for growth.

## New Zealand shorelines

The shores of New Zealand do not have the thick growths of wracks found in the north. Instead, there are far more barnacles and mussels on the upper and middle shores. Below these, there is a band of the brown seaweed Neptune's necklace (*Hormosira banksii*). Green seaweeds are more numerous, too, and species such as *Codium adhaerens* and *Caulerpa sedoides* grow well. On the lower shore the red seaweeds *Corallina officinalis*, *Gelidium*, and *Jania* grow, and in the shallow coastal waters, kelps such as *Macrocystis pyrifera* and *Lessonia* flourish. Other brown seaweeds such as *Sargassum* and *Xiphophora*, which are related to wracks, also grow there.

## Rock pools

Rock pools give a good idea of conditions on the seabed below low tide. A fascinating variety of the smaller seaweeds provides shelter for an equally amazing number of sea creatures – anemones, crabs, shrimp, and small fishes. These pools are left behind as the tide goes out. Pools on the upper reaches of the beach may only occasionally be covered by salt water at very high tide. Rain water tops up the water level, and so they are less salty than those lower down the beach, and contain different seaweeds. The lower pools support seaweeds that do not like to be exposed to the air.

# ROCKY TROPICAL SHORES

The seashores of the tropics are divided into zones, just the same as those in cooler parts of the world. However, the seaweeds and shore animals have to be able to tolerate brighter light and higher temperatures.

There are many more red seaweeds in all zones. On the upper shore these grow in sheltered cracks in the rock. Only blue-green algae can survive on exposed rock faces. On the middle shore, red and green seaweeds make a seaweed "turf." Low-growing tufts grow on bare rock or on encrusting seaweeds such as *Lithothamnion*. This algal "turf" continues down the shore to low water level. Brown seaweeds only appear in warm, shallow seas.

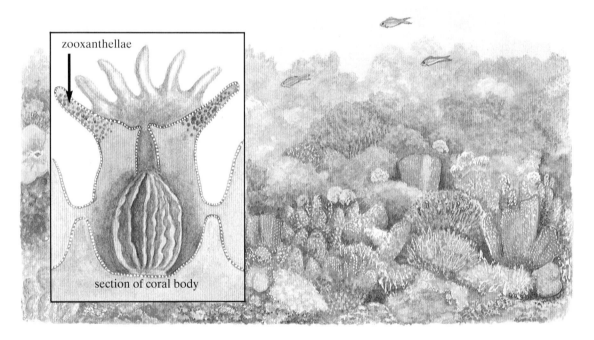

**Coral reefs**
Coral reefs are among the richest of the world's habitats. They are found only in warm, shallow, tropical seas where the water does not fall below 64°F. The reef is a chalky ridge made entirely by coral animals and calcareous (chalk-producing) red seaweeds. It is not only these red seaweeds that help the growth of the reef, for inside the body of the corals are single-celled algae called **zooxanthellae**. This is a mutually beneficial partnership as the algae get protection within the coral and provide the coral with food (see p. 21). The algae and coral animal together can grow faster than the coral can by itself. Because algae need sunlight, reefs can only grow in shallow water.

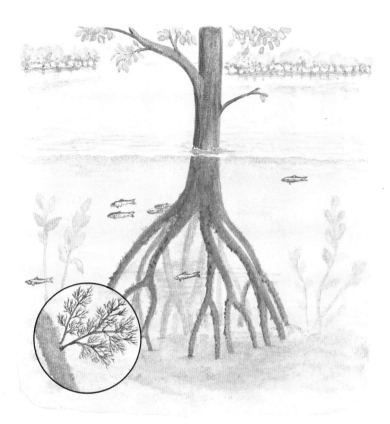

## Mangrove swamps

Sheltered, muddy bays or estuaries in the tropics have a special kind of vegetation. It is dominated by a group of trees called mangroves. They are different from all other trees because they are able to grow in shallow salt water. Mangroves have stilt-roots or kneelike projections that protrude from the water and absorb oxygen from the air. The underwater roots provide an anchorage for communities of seaweeds and smaller algae.

Red seaweeds are very numerous, and tufts of many different species grow in zones on the mangrove roots. Tangled among them are filamentous green (*Enteromorpha*), blue-green (*Lyngbya*), and red (*Callithamnion*) algae and chains of diatoms.

## Seaweeds of lagoons

A strange group of green seaweeds is found in the soft mud of mangrove swamps and sandy lagoons. These are the tubular or siphonaceous green seaweeds and include *Halimedia*, *Acetabularia*, *Caulerpa*, and *Udotea*. They are small to medium in size and each frond is made up of filaments. However, these are not divided into cells, so each filament behaves like a giant cell.

In sheltered bays of tropical Pacific islands, there are species of *Caulerpa* that produce poisons or toxins. Some fish are immune to these poisons and so they can eat the seaweeds. They in turn become so poisonous that it is very dangerous, sometimes fatally so, for people to eat them.

# LAKES

In some ways, large freshwater lakes are similar to oceans. They have a shore, which may be steep or gently sloping, sandy, stony, or muddy. There is only enough light for photosynthesis in the top 65 feet, maybe less, so no plants are found at the bottom of deep lakes.

Unlike the sea, there are no tides, although very large lakes like the Great Lakes of North America can have waves up to 20 feet high during storms. The rocks that make up the lake basin, and the rivers that flow into it, affect the water in the lake.

Freshwater lakes do not taste salty like the ocean. However, "fresh water" does not mean "pure" water. Chemicals from the rocks at the bottom of the lake dissolve in the water. Other chemicals are brought in by inflowing rivers, drainage from soil (run-off), and rotting plant and animal remains. These dissolved chemicals provide nutrients (food materials) for algae and higher (flowering) plants to grow. So lakes that are low in nutrients have fewer plants and algae than those with nutrient-rich water.

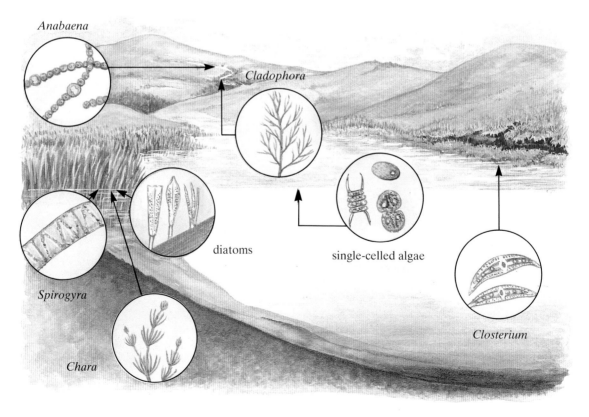

## Lakes in cold climates

Algae in Char Lake in the cold Canadian Arctic are limited to single-celled kinds such as desmids. The water is very low in the chemicals nitrate and phosphate, so the algae multiply slowly. Lake Myvatn in Iceland also has a cold climate, but this lake is fed by nutrient-rich springs. It supports many planktonic algae, and a thick carpet of blanket weed (*Cladophora*). These algae together provide food and shelter for hundreds of tiny water animals, which in turn feed enormous numbers of breeding wildfowl in spring and summer.

## Salt lakes

Some lakes have very high concentrations of minerals dissolved in them. These minerals, called "salts," do not necessarily taste at all like the salt you sprinkle on food or taste in seawater. Salts are brought to all lakes by inflowing rivers. They become concentrated in lakes in hot countries with little rainfall, and which have no outflowing river. Every year, large amounts of water **evaporate** from these lakes, leaving behind the salts. Often these salts dry into crystals around the edge of the lake.

The Great Salt Lake in Utah contains sodium chloride, the chemical name for common salt, but it is in much higher amounts than in seawater. Only one species of a single-celled green alga, *Dunaliella salina*, can live in its extremely salty water. Tiny brine shrimp feed on the algae.

# ALGAE OF OTHER HABITATS

Algae are found in reservoirs, rivers, and canals, as well as in lakes and oceans. They are also found in much smaller bodies of water – ponds, ditches, water barrels, and farmyard drinking troughs.

In running water, the smaller kinds live on the surface of stones and larger water plants. Filamentous algae often grow attached to rocks or pebbles. *Chlamydomonas* is often found in water barrels as well as ponds, and *Euglena* is common in stagnant water, which often turns bright green with huge numbers of this alga.

Many algae are found in the top layer of the soil. Diatoms, desmids, and blue-green algae grow well, often surviving dry seasons as tough **spores**, or resting stages. Large colonies of algae can be seen on the surface of moist soils. These colonies may be jellylike masses (*Nostoc*), felty skins (*Zygogonium*), or thin, greenish layers (*Mesotaenium*). In hot, tropical countries, blue-green algae help to stabilize sandy soils. Very large patches of *Nostoc commune* are grown in some parts of northern India, as they improve soil fertility.

**Green trunks?**
Algae that live out of water have to survive drought. One of the most widespread and successful of these is *Pleurococcus*, a bright green alga that lives on the bark of trees, damp fences, walls, and stones. It depends on rain and damp air for its supply of moisture and survives the summer as a dry, green powder.

## Algae as partners

Lichens are the result of algae and fungi living closely together as a partnership. The algae involved are various species of blue-green and green groups. They live among the hyphae or threads of the fungus. There are over 20,000 species of lichens.

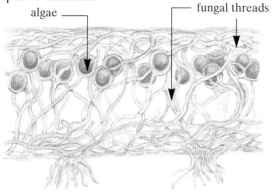

## Fern and algae

One species of blue-green alga, *Anabaena azollae*, lives in the rather spongy, floating leaves of the water fern *Azolla*. These are small ferns, growing mostly in the tropics. *Anabaena*, like some other blue-green algae, can trap nitrogen gas from the air and turn it into ammonium salts. The fern therefore has a built-in supply of nitrogen fertilizer. The Chinese have been growing *Azolla* and its lodger as a green manure for their rice fields for at least 2,000 years.

## Animals and algae

Tiny dinoflagellates, zooxanthellae, live inside the polyps of reef-building coral animals. The coral obtains a supply of sugars from the green algae. In return, it provides the algae with shelter and nitrogen-containing nutrients.

The green hydra (*Hydra viridissima*) is a small animal that lives in ponds and lakes. It is bright green because large numbers of green algae (chlorellae) live in its cells. Again, the hydra gets a supply of sugar from the algae.

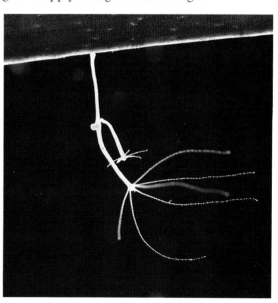

# SHAPE AND STRUCTURE

Single-celled algae are the most simple in structure, but even these display a most amazing variety of shapes. When seen under a powerful electron microscope, the cell walls of diatoms, desmids, and dinoflagellates are very complicated (see pp. 11, 13).

The larger seaweeds and stoneworts have a more clearly plant-like shape and some of them have quite a complex structure. Even the most advanced algae have no roots, leaves, flowers, fruits, or seeds like those found in flowering plants. Their appearances are so diverse that shape alone is not used by **botanists** (people who study plants) to sort them into groups. Instead, they are arranged, or classified, according to the pigment they contain. Of course, they all contain chlorophyll. In blue-green algae, blue pigment (phycocyanin) only partially hides the green chlorophyll.

Chlorophyll is, except in the blue-green algae, contained in tiny units called **chloroplasts**. It is here that photosynthesis takes place. The energy in sunlight triggers the chlorophyll molecule and starts off a chain of chemical reactions. Carbon dioxide gas and water combine to form sugars, oxygen, and water.

**Green algae**
Green algae do not contain large amounts of other colored pigments. Their cell walls are made of a compound called cellulose and food is stored in their cells as starch.

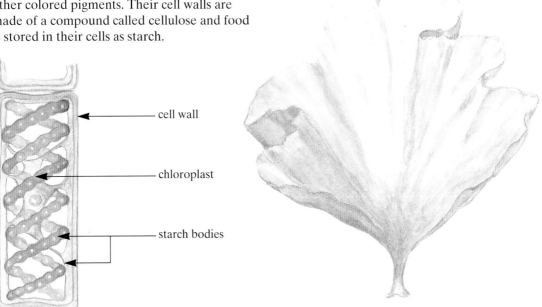

## Brown algae

In this group, the green chlorophyll is usually hidden by a brown-colored pigment called fucoxanthin. Food is stored as a sugary alcohol and carbohydrate, and the cell walls contain a lot of a substance called alginic acid (see p. 40).

There are no single-celled brown algae. Both the small, feathery seaweeds and the large wracks and kelps need hard surfaces to grow on. The thick, rubbery fronds of wracks and kelps can stand up to waves pounding them against rock, but not all of them can survive drying out. Kelps have a **holdfast** (a kind of root) and a thick stalk, or **stipe**, that can live for many years. The **fronds**, or blades, last only one year, and new blades grow from the top each spring.

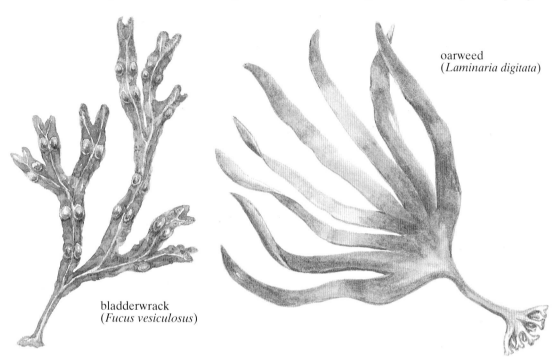

oarweed (*Laminaria digitata*)

bladderwrack (*Fucus vesiculosus*)

## Red algae

In this group, the green color is usually masked by a red pigment, phycoerythrin. Some have blue pigments, however, so "red" seaweeds may also be brown, blackish, or blue.

The chlorophyll of red algae is very slightly different from that of other algae. Sunlight is made up of different kinds of light. Some light rays have more energy than others and so pass deeper into the water. Red algae are able to use this light and so they can grow at greater depths than other seaweeds, and can flourish in the shade of kelp fronds.

Red seaweeds that grow on the shore tend to be wiry tufts with narrow, branched fronds that can resist wave damage. On the ocean floor, the water is calmer and species with wider, flatter fronds can grow.

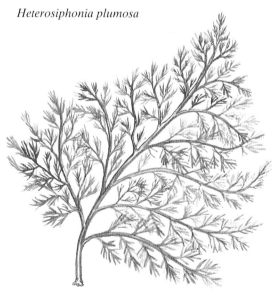

*Heterosiphonia plumosa*

# MOVEMENT AND GROWTH

Seaweeds are usually attached to rocks or other supports by a holdfast – they do not move about. However, single-celled and free-living colonies of algae are not fixed and many of them can move around. Some use tiny hairs called flagella but others, such as filaments of blue-green algae and tiny silica-encased diatoms, appear to move by magic. They do not have flagella, and the way they move is not yet known.

*Volvox* forms a colony of 500 to 50,000 individual cells, each of which is similar to *Chlamydomonas*, and has two flagella. A *Volvox* colony can be 1/50 of an inch across – clearly visible to the naked eye. Each colony is like a ball, with all the cells around the surface. The flagella lash in rhythm to rotate the tiny, green globe slowly through the water.

*Euglena* also has flagella, but instead of a stiff cell wall it has a flexible "skin" more like the cell wall of single-celled animals, called **protozoans**. *Euglena* has other animal-like features, including a digestive pouch or "gullet." As some *Euglena* contain chlorophyll, no one is sure if they are plants or animals.

**Flagella**
Many of these mobile algae have tiny flagella that they lash to propel themselves through the water. Some dinoflagellates found in plankton and a small group of green algae swim in this way.

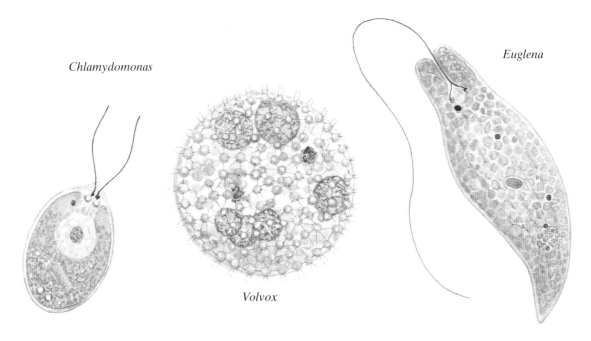

*Chlamydomonas*

*Volvox*

*Euglena*

## Growth

A full-grown, single-celled alga is still very tiny. Only a few, such as *Micrasterias* (a desmid), are large enough to be seen with the naked eye. So, in favorable conditions, these very tiny plants "grow" not by becoming larger, but by multiplying and increasing in numbers. This can happen suddenly – a stagnant pool becomes bright green with *Euglena*, a lake turns into pea soup with an explosion of *Gloeotrichia*, or *Gonyaulax* turns an incoming tide red.

Although as plants these algae make their own food by photosynthesis, they also need other nutrients. When all the nutrient supply is used up, rapid growth cannot continue and the numbers of algae fall as fast as they rose. The rise and fall of algal populations, called a "bloom," is part of the normal seasonal cycle of the ocean or a lake.

## Nitrogen Fixation

Nitrogen gas in the air does not react with other chemicals easily. Plants and animals cannot live without nitrogen, but it has to be in a form, such as nitrates or ammonium, that they can use. Some blue-green algae such as *Anabaena* and *Nostoc* can "fix" nitrogen gas. They have special cells called **heterocysts** that can trap the nitrogen and release it as ammonium, which can then be used by both algae and flowering plants.

heterocysts

# LIFE CYCLES OF ALGAE

Many algae produce new plants by simply dividing to make two new cells called daughter cells. Filaments of green and blue-green algae spread by **fragmentation** in which short lengths of filament break off and grow into new plants. Sometimes, the contents of an algal cell reorganize to form a **zoospore**. Special cells along the filaments of *Oedogonium* each release a single zoospore, which has a ring of flagella near one end and thus can swim through the water. The cell contents of *Chlamydomonas* divide to make two, four, or eight mobile zoospores. These zoospores grow into new plants. The process of making new plants in these ways is known as vegetative reproduction.

When a cell divides to give two daughter cells, each one is identical to the other. All the offspring formed in this way will also be identical. When two unrelated algae unite to produce new individuals, it is known as **sexual reproduction**. It gives the chance of slight variations, and keeps the species healthy. In simple plants and animals, there is often very little difference between "male" and "female" partners. Instead of producing seeds, as flowers do, many algae make spores that grow into the next generation.

**Vegetative reproduction**
Single-celled algae like desmids reproduce by dividing into two cells. Filamentous algae like *Hormidium* spread when short lengths break off and grow into new plants.

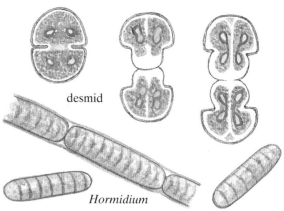

desmid

*Hormidium*

*Spirogyra*
The cells of two filaments of *Spirogyra* lying next to one another grow small swellings. These eventually join to form a connecting tube. The contents of one cell pass through this tube and mix, or fuse, with the contents of the other cell. A dark, roundish spore is formed. Its thick wall helps it to survive conditions such as drought or extremes of temperature. When better conditions return, the spore splits and a new filament emerges.

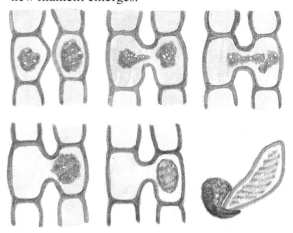

## Seaweeds

Seaweeds have complicated life cycles with two or three frond stages. The parent plant of a seaweed is called a **sporophyte** and the other frond stages **gametophytes**. The sea lettuce (*Ulva lactuca*) has a sporophyte and a gametophyte that look exactly the same.

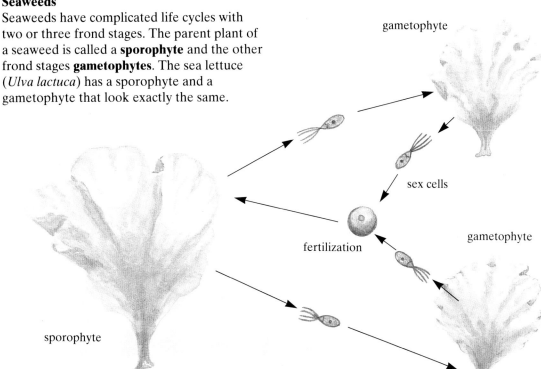

## Oarweed *(Laminaria digitata)*

This large kelp produces mobile spores. When these settle on suitable rocks, they each grow into either a small male or a small female plant. These are the gametophytes. They are quite unlike the large, tough kelp. Mobile male cells are released from the male plant. They are attracted to the female plants by special chemicals, where they **fertilize** the much larger female egg cells. These fertilized cells grow into a new oarweed.

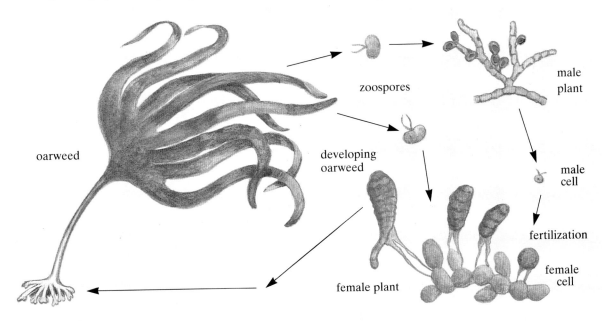

# GRASS OF THE SEA

Plankton grow best in nutrient-rich water but, because light is essential, they only grow in the top layer of the ocean. The nutrients here are soon used up unless they are continually replaced. Close to the shore, water run-off from the land and river outflows supply many nutrients. Out at sea, however, algae depend on food materials brought up to the surface by the mixing of deeper waters with the upper layers. In tropical seas, the surface layer of the water is warm and therefore lighter, so it floats on the colder water. There is little mixing. The rate of algal growth in the tropics is therefore low, although it continues steadily all year.

In temperate and polar seas, algal growth is seasonal. During the long, polar winters, growth is limited by lack of light. In the North Atlantic, there are two peaks of growth. In spring, the surface layer is well supplied with nutrients after the winter storms and the numbers of algae increase. During the summer, the surface layer of water warms up

**Life in the North Atlantic**
The life cycle of marine animals is closely linked with the growth of plankton and the tiny animals that feed on them. Cod, haddock, and whiting, the commercially important fish of the North Atlantic, lay their eggs in the spring. Newly hatched fish (fry) live at the surface, feeding on the plankton. As they grow larger, they swim to the seabed. Herrings, however, stay near the surface, feeding on plankton.

and no longer mixes with the heavier, colder layers. The nutrients are used up and the numbers of algae fall. (The separation of water into these layers is called a **thermocline**.) In autumn the temperature falls, and the surface water cools and sinks. The thermocline breaks down, and more fertile waters come to the surface, causing the autumn peak of growth.

In the southern hemisphere, there is less land. Cold currents of water flow around Antarctica all year, continually churning up the water. So the growth of algae continues at a high rate throughout spring, summer, and autumn, until the onset of winter when it is too dark for photosynthesis.

**A fertile shore**
Along the coast of Peru, the cold Humboldt current meets the continental shelf. The deep water wells up to mix with the warm upper layers of water. Planktonic algae flourish in the constant supply of nutrients. The water teems with fish, and there is a thriving fishing industry.

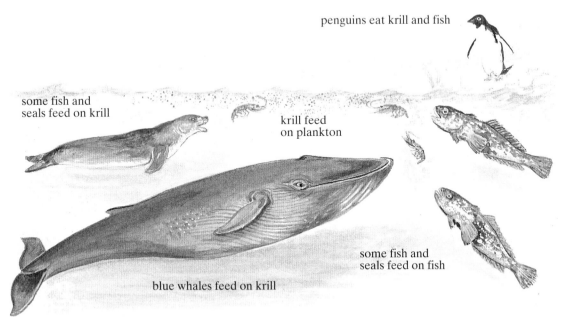

penguins eat krill and fish
some fish and seals feed on krill
krill feed on plankton
some fish and seals feed on fish
blue whales feed on krill

**Antarctic food webs**
In the southern ocean around the Antarctic, a 2 inch-long shrimp, *Euphausia superba*, feeds on the planktonic algae. These shrimps are known as krill. The largest animal that has ever lived, the blue whale, feeds all summer on krill, filtering it from huge mouthfuls of seawater through plates of baleen. Millions of tons of krill are eaten every year by fishes, seabirds, and seals, as well as whales.

# THE KELP FOREST

Beds of kelp are widespread, but the largest kelp "forests" are found in the southern hemisphere. The exception are the beds of giant kelp growing off the shores of California. The kelp beds that grow off the west coast of southern Africa are especially large, stretching as far as two miles offshore. They are fed by the rising waters of the cold Benguela current.

Kelp forests can be compared with forests of woodland trees. They have a canopy, an understory, and a forest floor. In a forest of trees, new tree seedlings can only grow when a large, old tree falls, allowing light in through the canopy. Similarly, young kelp plants cannot grow in the deep shade of the larger kelps. When these are torn off by violent waves, the young kelps spring up.

### Giant kelp (*Macrocystis pyrifera*)

This brown seaweed is one of the world's largest plants – it can grow to over 400 feet in length. It is found off many coasts in the southern hemisphere, and also along the Pacific coast of North America. Like all kelps, it does not grow in tropical seas.

It grows very quickly – up to 12 inches a day. The long fronds, which have many lobes, grow from the top of a thick stem. At the base of each lobe is an air bladder, which keeps the frond floating near the surface.

### Camouflage

Seaweed sometimes provides shelter and protection for animals. A relative of the sea horse, the leafy sea dragon, lives in the kelp forests of southern Australia. It looks just like a tattered piece of kelp frond so it blends in with its background. Pipe fishes also conceal themselves in seaweed, feeding on plankton.

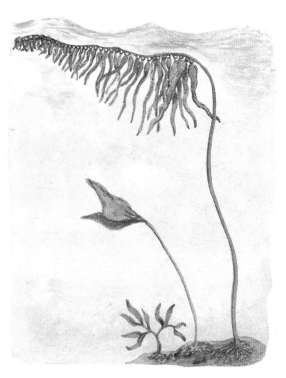

**Sea bamboo**
***(Ecklonia maxima)***
The main alga in the kelp beds of southern Africa is the sea bamboo, which has a hollow, gas-filled stipe up to 40 feet long. Just below the handlike frond, the stipe swells to make a float that keeps the frond near the surface of the water.

**Animals of Kelp Forests**
Few animals graze on the living kelp plants. The fronds are slippery and hard to cling to as the strong waves fling them to and fro. Sometimes the animals that depend on them do not live in the kelp bed at all. Fronds that are washed ashore support large communities of tiny animals like sandhoppers and isopods (small crustaceans). Pieces of kelp also sink to the bottom and bacteria (simple microscopic organisms) feed on the kelp debris. Over 70 percent of animals that feed on kelp are filter-feeders, living on tiny fragments in the seawater. Sea urchins are also common on the floor of the kelp bed. They eat fragments of kelp and the young kelp plants, and lobsters feed on the sea urchins.

In the forests of giant kelp off the southern coast of Australia and Tasmania, fur seals hunt among the kelp stems, feeding on fish, squid, and prawns. The fur seals are themselves preyed on by the great white shark.

# LIFE IN FRESH WATER

On pages 18 and 19 you saw how different communities of algae occupy different parts of the lake. Wherever these lakes are in the world, there are animals to take advantage of this source of food. Food chains in lakes and rivers may be short, or they may have many links (see below).

The largest group of freshwater animals that depend on algae are the fish, which range from tiny, jewel-like tropical tetras to large, fierce predators such as the pike and the huge muskellunge of North America. The pochard, a duck that lives in the U.K., Europe, and Asia, feeds mainly on stoneworts (*Chara, Nitella*, and *Tolypella*). Algae are a major part of the diet of other plant-eating ducks, including many species of pintails.

The **larvae** (young forms) of many flying insects live in fresh water. Some feed on debris, while others eat algae or other animals. Stoneflies, mayflies, caddis flies, midges, mosquitoes, and dragonflies all form part of the food chain. The adults of each kind emerge at about the same time. Some live just long enough to mate and lay eggs. Many animals feed their young on these insects.

Bats are attracted to swarms of insects. Daubenton's bat and the little brown bat in particular hunt over water.

Swifts, swallows, martins, and wagtails are groups of insect-eating birds that hunt over water.

Dragonflies and damselflies hide in waterside vegetation while hunting other insect prey.

**Frogs and toads**

Amphibians such as frogs, toads, newts, and salamanders are all able to live on land, but they have to return to water to spawn. Their young are tadpoles – quite unlike the adult animal. At first, tadpoles feed by nibbling off the layer of algae covering plants and pebbles. As they grow larger, they become carnivorous, eating other water animals, dead or alive. Frogs and toads prefer smaller, shallow ponds. They breed very successfully in garden pools or in swamps, where they can be observed in their natural habitat.

**Soda lakes of East Africa**

There are a number of lakes in the eastern rift of Africa that contain high levels of the chemical sodium carbonate. In Kenya, Lake Nakuru is the most well known and is now a national park. The only plant that is able to grow in the soda-rich waters is one species of blue-green algae, *Spirulina platensis*. This alga grows in great quantities and provides food for one kind of copepod (a small water animal), one species of small cichlid fish, and the lesser flamingo. Only a few species of insects and rotifers (small water animals) feed on the copepods, but there are so many insects and fishes in the lake that large flocks of many different kinds of birds also come to feed.

# THREATS TO LAKES

Shallow, lowland lakes in areas with a fertile soil are naturally nutrient-rich, or **eutrophic**. Thick vegetation fringes their shores and water plants flourish. The water in deep lakes, often found in mountains, contains little oxygen and is naturally low in nutrients and organic material. These lakes contain particular plants and animals.

More lakes are becoming eutrophic. The amount of nutrients they contain, especially nitrates and phosphates, is increasing. These chemicals come partly from fields where farmers use fertilizers to obtain higher yields from their crops. The chemicals soak through the soil and eventually drain into rivers and lakes. Because people like to live beside lakes, many towns, cities, and residential areas expand and lakes all too often become overburdened with waste water and sewage.

**Algae as indicators of pollution**
The amount of pollution can often be assessed by identifying the algae living in the water. Samples are taken from the water and sediments on the bottom. If the water changes in any way, for example the phosphate level rises, the types of algae found also change. Some species become more scarce, while others flourish, so algae are useful and sensitive indicators of pollution.

Heavy pollution with poisonous chemicals: nothing grows

Extremely heavy pollution with raw sewage: bacteria only

Heavy sewage pollution: bacteria (*Thiothrix*)

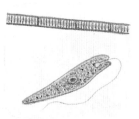

Less heavy pollution, but still no oxygen: *Euglena, Oscillatoria*

Some oxygen in water: *Ulothrix*

Eutrophic water, natural or artificially enriched: *Cladophora*

Unpolluted water: *Batrachospermum*

Water with very few dissolved minerals: *Draparnaldia*

Each community contains a number of species. Only the most numerous or abundant are shown.

## Polluted waters

People today have more leisure time and many have taken up boating. However, increasing numbers of small boats are ruining many beauty spots. Gasoline escaping from outboard motors poisons the surface film of water. The action of the propellers stirs up mud from the bottom. This makes the water cloudy and turbid, so there is not enough light. Oxygenating plants die, and the water becomes stagnant.

In the 1950s and 1960s, the people of Seattle became increasingly aware that waste from the city was polluting Lake Washington. In the summer months, blue-green algae such as *Oscillatoria* thrived on the increasing amounts of phosphorus in the water. During the 1960s, the sewage was diverted away from the lake and discharged far out at sea. Since then, the phosphate levels have fallen and the lake has made a good recovery.

### Phosphates

Fresh water usually contains low levels of phosphates. The use of detergent washing powders, which contain a lot of phosphate, is high in wealthy nations and the waste water often drains into lakes and rivers. This upsets the balance of nutrients in the lake and the extra phosphate causes rapid growth of blue-green algae resulting in many problems:

- The algae make a scum on the water surface.
- Algae prevent light from penetrating the water, so water plants die.
- As plants and algae die, their decay uses up all the oxygen in the lake. Then animals die, and that uses up more oxygen.
- Water becomes stagnant, smelly, and foul.
- Blue-green algae are often toxic, poisoning fish as well as animals drinking the water.

# THREATS TO MARINE LIFE

The list of pollutants that are dumped in the ocean is a long one. Concerned people, and some countries and governments, are trying to get an international agreement to stop waste from being dumped into the ocean. However, the pressures to dispose of dangerous waste are very great in some countries, and not all countries have suitable means to dispose of such waste. Also, it is often difficult for many different countries to reach an agreement.

**Chemicals Dumped at Sea**

What happens to poisonous chemicals when they are in the ocean? There is plenty of water to dilute them, but this does not mean that they disappear. Some sink to the ocean floor where they are swallowed by fish and crustaceans and so enter the food chain. Others are taken up by planktonic algae and animals. Poisons become more concentrated higher up the food chain, so the top predator has the highest dose. The most common chemicals dumped are given here:

DDT – banned in North America and Europe, but still used in undeveloped countries. Dieldrin, Aldrin, Endrin – highly toxic pesticides, similar to DDT.
PCBs – polychlorinated biphenyls, used in the electronics industries, are burned out at sea.
Heavy metals – mercury, lead, zinc, cadmium come from many industrial sources.
Radioactive materials – small amounts come from nuclear power stations. High-level radioactive wastes are sealed in lead-lined concrete, or glass blocks.

## Oil

About 1.5 billion gallons of oil get into the ocean every year. Many large oil spills are accidental and they attract a lot of publicity, but they only add up to about 4 percent of the total. Oil, and the detergent used to clean it up, kills seaweed and plankton and harms seabirds and shoreline animals such as sea otters.

## Sewage

Sewage is discharged into the ocean, often close to the shore. It contaminates the water and beaches. This is not only unpleasant, but also a health hazard as raw sewage contains pathogenic (disease-causing) bacteria and eggs of parasites that live in the human gut. As in fresh water, the large amounts of nitrates and phosphates cause the algae to multiply so rapidly that they can poison the fish. They also screen out sunlight so that the animals and plants die.

Land-enclosed seas suffer the most. Only a small amount of water enters or leaves the Mediterranean Sea, through the narrow Straits of Gibraltar, into the Atlantic Ocean, and pollutants become trapped. The area is crisscrossed with oil tanker routes and the water is polluted both from accidental spills and the deliberate, although illegal, habit of washing out empty oil tanks at sea before refilling.

Many people live around the Mediterranean and Adriatic seas, and tourists flock there in summer, so sewage pollution is also a serious problem. Vacationers do not want to swim in water thick with slimy algae that thrive in the warm, nutrient-enriched waters.

# RESERVOIRS

People in developed countries usually have an unlimited supply of clean water. Every day people and industry use a great deal of water, most of which comes from large lakes. In places where there are no lakes, or where the lakes are too small, reservoirs are built to collect and store water. Even in countries with fairly high rainfall, like Britain, large cities use more water than natural resources can supply.

Reservoirs are usually flooded river valleys. A dam is built and the water collects behind it. The original vegetation of the valley is flooded, and an artificial lake is formed. Some reservoirs are built by people specifically to store water.

**Pollution in reservoirs**
Industrial wastes and raw sewage cannot be discharged into reservoirs and motor boats are forbidden. However, nitrates and phosphates can enter reservoir water from treated sewage and fertilizer run-off (see p. 34). Nitrates are difficult, and expensive, to remove, but strict limits are being set on the nitrate levels in drinking water, for too much may be unsafe.

Europe's largest reservoir for water storage is Rutland Water, in the U.K. During the warm, dry summer of 1989, the water level dropped considerably. Hot weather and high levels of nitrates and phosphates encouraged a massive algal bloom of *Microcystis*. Some dogs and sheep that drank the water were killed.

## Waterworks

Planktonic and filamentous algae soon colonize reservoir water. In a deep, well-oxygenated reservoir in an upland area, water can be drawn from the lower levels where algae cannot grow. However, algae do develop in lowland and water-storage reservoirs with a high level of nutrients. These algae have to be filtered out.

Water drawn from a reservoir first passes through filter beds.
Coarse filters remove larger filamentous algae, large planktonic animals such as water fleas (*Daphnia*).
Fine filters remove smaller algae and most bacteria.

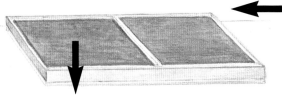

A growth of algae develops on the surface of the fine filters. These algae help to remove bacteria and oxygenate the water as it passes through.

The filtered water is chlorinated to kill any remaining bacteria and algae. Other chemicals are added to remove unwanted color and dissolved organic substances. People are concerned that some additives such as aluminum hydroxide damage health.

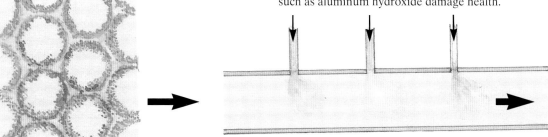

## Algae in sewage treatment

Sewage is treated before being discharged to kill harmful organisms and to break down solid matter so that it spreads more readily in the water. During treatment, sewage is passed into large, shallow pans. The sewage is kept moist and algae develop over the surface. Oxygen is plentiful as green algae, such as *Chlorella* and *Scenedesmus*, flourish. These are important in both temperate and tropical regions for the rapid breakdown of sewage.

# SEAWEED AND INDUSTRY

A very useful substance, alginic acid, can be obtained from large brown seaweeds such as kelps and wracks. Alginic acid occurs in the cell walls of these plants, and makes up between 14 and 40 percent of their dry weight.

Harvesting the seaweed can be difficult. Some are gathered from rocky shores from small boats or by hand. The largest enterprises harvest giant kelps such as *Macrocystis*. These are cut by machine and gathered into large boats. Kelps grow very quickly – as much as 10 to 15 feet in a week. So the kelp plants soon regrow.

Carrageenin can be extracted from red seaweeds, mainly Irish moss (*Chondrus crispus*) and *Gigartina*. *Chondrus crispus* grows well along the shoreline of Canada and it is gathered with large wooden rakes when the tide is out. Carrageenin has very similar uses to alginates (see the panel opposite).

**Agar**

Agar is purified from red seaweeds. These small plants are even more awkward to harvest than kelps. Most are gathered by hand. The species used come from many parts of the world.

Purified agar is a powder that dissolves in hot water. It sets to form a jelly. This has proved to be invaluable for culturing, or growing, bacteria and molds. It is used in hospitals, laboratories, and industries worldwide.

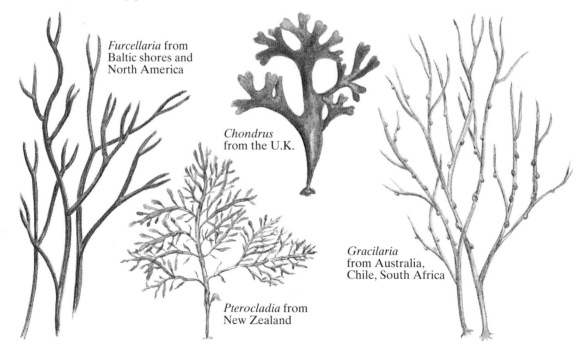

*Furcellaria* from Baltic shores and North America

*Chondrus* from the U.K.

*Pterocladia* from New Zealand

*Gracilaria* from Australia, Chile, South Africa

## Processing Seaweed

Chemicals called "alginates" can be made from alginic acid. They are nontoxic, and can be made into a thick paste or jelly. They have many uses:

■ Food industry: in cream fillings, ice cream, sweets, jelling agent in meat products, processed cheese.

■ Drugs and cosmetic industries: used in medicines, cosmetic and medical creams, made into fibers used for medical gauze.

■ Other industries: printing pastes for textiles, manufacture of latex rubber, surface finishes for paper.

■ Species used: *Ascophyllum* (northern coasts); *Laminaria* (northern coasts and South Africa); *Macrocystis* (California, Australia, and South America); *Ecklonia* (Australia, Japan, and South Africa); *Eisenia* (Japan); *Lessonia* and *Durvillea* (South America).

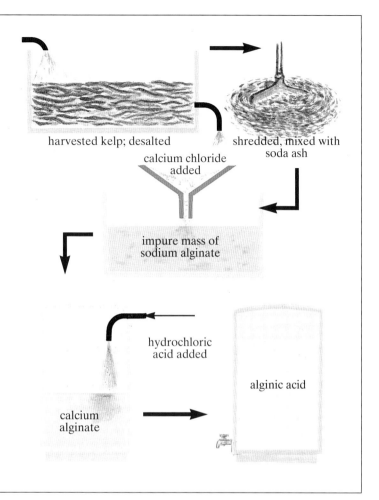

## Diatomite

When diatoms die, they sink to the bottom of the water. Their silica shells do not decay. Although tiny, diatoms are so numerous that, with the passing of years, diatom shells form a thick sediment. Fossil deposits, known as diatomite or kieselguhr, are found in sites of ancient seabeds such as Lompoc in California. There are also freshwater deposits of diatomite, for instance in Kentmere, U.K.

Diatomite can be as much as 88 percent silica. It does not react easily with other chemicals and does not burn. It has a number of industrial uses such as a filler for paper, insulation products, paints, and industrial abrasives. It is also used as a filter in many industrial processes, such as sugar-refining, wine and brewing industries, and chemical industries.

# THE FUTURE

Many people in coastal communities have eaten seaweed and some still do. *Porphyra* is eaten in west Wales, where it is made into laver bread, and in Ireland. Dulse (*Rhodymenia*) is also part of traditional Irish fare. Seaweed is far more important in the diet of the people of the Philippines, Indonesia, Pacific Islands, the coast of China, and especially Japan.

Algae may be used in many ways in the future. One new idea being developed in South Africa is the use of kelp extract to stimulate, or increase, the growth of crops. The growth rates of wheat treated with kelp that had been made into a liquid showed an improvement by 20 to 300 percent.

In the U.S., the possibility of producing methane (natural gas) from kelp, taking advantage of the fast rates of growth of these seaweeds, is being studied. Kelp is grown on huge racks underwater, and nutrient-rich water is pumped up from deeper levels. If successful, this could be a breakthrough in the search for a renewable energy source.

### Spirulina
This blue-green alga grows in salt and soda lakes of Mexico and parts of Africa. It is the only plant that can survive in many of these lakes, and forms a matlike growth on the water surface. This can easily be skimmed off. In Mexico, *Spirulina* is now added to chicken food, but the Aztecs once ate sun-dried "wafers" made from this alga. It is rich in protein and vitamin $B_{12}$, and it grows very quickly. It may be a nutritious addition to the often-poor diet of people who live in the poorer countries of Africa.

## Japanese seaweed farms

Thousands of tons of seaweed are harvested every year by the Japanese, who also cultivate the seaweed on special farms. *Porphyra* "amanori," *Laminaria* "kombu," and *Monostroma* "laonori" are three of the most popular seaweed varieties. They are grown on bundles of bamboo stuck into the mud or on nets made of palm fibers. The seaweed spores settle and grow. Nets of the young plants can be moved to the best positions in the shallow waters of bays and estuaries. Seaweeds are rich in sugars, starches, vitamins A and C, and are a good source of iodine.

## Animal fodder and fertilizers

Seaweed and algae are sometimes added to animal fodder or food. Seaweed in particular is a good source of trace elements, nutrients needed by animals but only in small amounts.

Seaweeds rot quickly giving a rich, natural compost. Farmers living near the coast have used it for centuries, and gardeners continue to take advantage of seaweeds thrown up by the tides to enrich their soil. A liquid extract of seaweed is produced commercially. Seaweed fertilizers contain high levels of potassium and crushed calcareous red seaweeds are also used on soils lacking lime.

# GLOSSARY

BIOLUMINESCENCE – The light made by some living organisms.
BOTANISTS – People who study plants.
CHLOROPHYLL – The green pigment that uses energy in sunlight to make food by photosynthesis.
CHLOROPLASTS – Tiny bodies in the cells of green plants that contain the pigment chlorophyll.
EUTROPHIC – Water that is rich in organic nutrients.
FERTILIZE – The fusing of a male cell with a female to form a new individual.
FILAMENT – A long thread made up of a single row of algal cells.
FLAGELLA (singular flagellum) – The minute hairs that enable some algae to swim through the water.
FRAGMENTATION – This is when short lengths of algal filaments break off and grow into new plants.
FROND – The algal plant except for the holdfast.
GAMETOPHYTE – The stage of some algae that produces the sex cells.
HETEROCYSTS – Special cells in blue-green algae where nitrogen gas is trapped and converted into nitrogenous compounds.
HOLDFAST – The tough base of a seaweed that attaches it firmly to a rock or other seaweed.
LARVAE (singular larva) – The young stages of invertebrate animals, often very different from the adult.
LITTORAL ZONE – The part of the seashore covered and uncovered by the tides.
PHOTOSYNTHESIS – The process in chlorophyll-containing plants that uses sunlight to convert carbon dioxide gas and water into sugars, releasing oxygen.
PLANKTON – Minute, free-living plants and animals that live in open waters.
PROTOZOANS – Single-celled animals.
SEXUAL REPRODUCTION – The means by which living things increase in numbers by combining sex cells from different individuals.
SPORES – Very tiny units produced by nonflowering plants, including algae, that grow into new individuals. They are often a means of dispersal or survival during unfavorable conditions such as droughts.
SPOROPHYTE – The stage of some algae that produces spores.
STIPE – The tough stalk of a seaweed.
THERMOCLINE – The separation of lake and salt water into cold and warm layers.
ZOOSPORE – A spore that has flagella and so can swim through the water.
ZOOXANTHELLAE (singular zooxanthella) – Single-celled, golden-brown algae that live inside corals and other marine animals.

## FURTHER READING
**For children:**
*Green Magic: Algae Rediscovered* by Lucy E. Kavaler; Crowell Jr. Bks., 1983.
*Protecting the Oceans* by John Baines; Steck-Vaughn, 1990.
**For adults:**
*Introduction to the Algae* (2d ed.), by Harold C. Bold and Michael J. Wynne; Prentice Hall, 1985.
*Spirulina: Food for a Hungry World; A Pioneer's Story in Aquaculture* by Hiroshi Nakamura; Univ. Trees, 1982.

# ALGAE IN THIS BOOK

Acetabularia
Anabaena
Anabaena azollae
Ascophyllum
Batrachospermum
Bladderwrack (*Fucus vesiculosus*)
Blanket weed (*Cladophora*)
Callithamnion
Caulerpa
Caulerpa sedoides
Ceratium
Chara
Chlamydomonas
Chlorella
Chorda filum
Chroomonas
Closterium
Codium
Codium adhaerens
Corallina
Corallina officinalis
Coscinodiscus
Draparnaldia
Dulse (*Rhodymenia*)
Dunaliella salina
Durvillea
Ecklonia
Ectocarpus
Eisenia
Enteromorpha
Euglena
Furcellaria lumbricalis
Gelidium
Giant kelp (*Macrocystis pyrifera*)
Gigartina
Gloeotrichia
Gonyaulax
Gracilaria verrucosa
Gymnodinium brevis
Halimedia
Heterosiphonia plumosa
Hormidium
Irish moss (*Chondrus crispus*)
Jania

Laminaria
Lessonia
Lithothamnion
Lyngbya
Mesotaenium
Micrasterias
Microcystis
Monostroma
Neptune's necklace (*Hormosira banksii*)
Nitella
Nostoc
Nostoc commune
Oarweed (*Laminaria digitata*)
Oedogonium
Oscillatoria
Pediastrum
Pinnularia
Pleurococcus
Porphyra
Pterocladia capillacea
Ralfsia
Sargassum
Scenedesmus
Scytosiphon
Sea bamboo (*Ecklonia maxima*)
Sea lettuce (*Ulva lactuca*)
Spirogyra
Spirulina
Spirulina platensis
Sugar kelp (*Laminaria saccharina*)
Synechococcus
Tolypella
Udotea
Ulothrix
Ulva lactuca
Volvox
Xiphophora
Zygogonium

# INDEX

**A**
*Acetabularia* 17
agar 40
alginates 40, 41
alginic acid 23, 40, 41
amphibians 33
*Anabaena* 9, 18, 25
*Anabaena azollae* 21
Antarctic 29
Arctic 19
*Ascophyllum* 41
Atlantic coasts 14
*Azolla* 21

**B**
bacteria 10, 31, 34, 37, 39
*Batrachospermum* 34
Benguela current 30
bioluminescence 13
bladderwrack 23
blanket weed 19
"bloom" 25
blue-green algae 9, 10, 16, 17, 20, 21, 22, 24, 25, 26, 33, 35, 42
brown algae 10, 11, 15, 16, 23, 30, 40

**C**
*Callithamnion* 17
camouflage 30
carrageenin 40
*Caulerpa* 17
*Caulerpa sedoides* 15
cellulose 22
*Ceratium* 11
*Chara* 9, 18, 32
chemicals 18, 34, 36
*Chlamydomonas* 11, 20, 24, 26
*Chlorella* 39
chlorellae 21
chlorophyll 8, 9, 10, 22, 23, 24
chloroplasts 22
*Chondrus crispus* 40-41

*Chorda filum* 11
*Chroomonas* 11
*Cladophora* 11, 18, 19, 34
*Closterium* 9, 18
*Codium* 11
*Codium adhaerens* 15
cold lakes 19
continental shelf 29
*Corallina* 10
*Corallina officinalis* 15
coral reefs 16, 21
*Coscinodiscus* 11
cryptomonads 11

**D**
daughter cells 26
desmids 19, 20, 22, 25, 26
diatomite 41
diatoms 10-13, 17, 18, 20, 22, 24, 41
dinoflagellates 10-13, 21, 22, 24
*Draparnaldia* 34
dulse 42
*Dunaliella salina* 19
*Durvillea* 41

**E**
Earth, formation of 9
*Ecklonia* 41
*Ecklonia maxima* 31
*Ectocarpus* 11
*Eisenia* 41
energy source 42
*Enteromorpha* 17
*Euglena* 11, 20, 24, 25, 34
*Euphausia superba* 29
eutrophic lakes 34

**F**
ferns 21
filaments 8, 10, 17, 24, 26
flagella 11, 24, 26
food chains 12, 32-33

fossils 9
fragmentation 26
fronds 23, 30, 31
fucoxanthin 23
*Fucus vesiculosus* 23
fungi 21
*Furcellaria lumbricalis* 40

**G**
gametophyte 27
*Gelidium* 15
giant kelp 8, 30, 31, 40
*Gigartina* 40
*Gloeotrichia* 25
*Gonyaulax* 13, 25
*Gracilaria verrucosa* 40
Great Salt Lake 19
green algae 10, 11, 15, 16, 17, 19, 20, 21, 22, 24, 26, 39
*Gymnodinium brevis* 13

**H**
*Halimedia* 17
heterocysts 25
*Heterosiphonia plumosa* 23
holdfast 23, 24
*Hormidium* 26
*Hormosira banksii* 15
Humboldt current 29
*Hydra viridissima* 21

**I**
industrial uses 40-41, 43
insect larvae 32
Irish moss 40-41

**J**
*Jania* 15

**K**
kelps 23, 27, 30-31, 40, 41, 42
kieselghur 41
krill 29

46

## L

lagoons  17
*Laminaria*  41, 43
*Laminaria digitata*  23, 27
*Laminaria saccharina*  11
*Lessonia*  15, 41
lichens  21
*Lithothamnion*  16
littoral zone  14
*Lyngbya*  17

## M

*Macrocystis pyrifera*  8, 15, 30, 31, 40, 41
mangroves  17
manure  19, 43
*Mesotaenium*  20
methane  42
*Micrasterias*  25
*Microcystis*  10, 38
minerals  19, 34
*Monostroma*  43

## N

Neptune's necklace  15
New Zealand shorelines  15
*Nitella*  32
nitrates  19, 34, 37, 38
nitrogen fixation  21, 25
*Nostoc*  20, 25
*Nostoc commune*  20
nutrients  18-19, 25, 28-29, 34, 35

## O

oarweed  23, 27
*Oedogonium*  26
oil  37
*Oscillatoria*  10, 34, 35

## P

Pacific islands  17
*Pediastrum*  11
phosphates  19, 34, 35, 37, 38
photosynthesis  8-9, 12, 18, 22, 25, 29
phycocyanin  22
phycoerythin  23
pigment  8-9, 13, 22-23
*Pinnularia*  11

plankton  10, 12, 19, 24, 28-29, 30, 37
*Pleurococcus*  20
poisons  13, 17, 35, 36
polar seas  28
pollution  34-39
*Porphyra*  10, 42, 43
*Pterocladia capillacea*  40

## R

*Ralfsia*  11
red algae  10, 15, 16, 17, 23, 40
reproduction  26
*Rhodymenia*  8, 42
rock pools  15
run-off  18, 28, 34, 38

## S

salt lakes  19, 42
*Sargassum*  8, 14, 15
*Scenedesmus*  39
*Scytosiphon*  11
sea bamboo  31
sea lettuce  11, 27
seaweed farms  43
sewage  34, 35, 37, 38, 39
sexual reproduction  26
soda lakes  33, 42
*Spirogyra*  11, 18, 26
*Spirulina*  42
*Spirulina platensis*  33
spores  20, 26, 27, 43
sporophyte  27
stipe  23, 31
stoneworts  10, 22, 32
sugar kelp  11
*Synechococcus*  10

## T

thermocline  29
*Thiothrix*  34
*Tolypella*  32
toxins  13, 17, 35, 36
tropical seas  16-17, 28

## U

*Udotea*  17
*Ulothrix*  34
*Ulva lactuca*  11, 27

## V

vegetative reproduction  26
*Volvox*  8, 24

## W

wracks  14, 15, 23, 40

## X

*Xiphophora*  15

## Z

zoospore  26
zooxanthellae  16, 21
*Zygogonium*  20

47